SpringerBriefs in Applied Sciences and Technology

More information about this series at http://www.springer.com/series/8884

Nuno M.M. Ramos · João M.P.Q. Delgado
Ricardo M.S.F. Almeida
Maria L. Simões · Sofia Manuel

Application of Data Mining Techniques in the Analysis of Indoor Hygrothermal Conditions

 Springer

Nuno M.M. Ramos
Department of Civil Engineering
University of Porto
Porto
Portugal

Maria L. Simões
Department of Civil Engineering
University of Porto
Porto
Portugal

João M.P.Q. Delgado
Department of Civil Engineering
University of Porto
Porto
Portugal

Sofia Manuel
Department of Civil Engineering
University of Porto
Porto
Portugal

Ricardo M.S.F. Almeida
Department of Civil Engineering
Polytechnic Institute of Viseu
Viseu
Portugal

ISSN 2191-530X   ISSN 2191-5318   (electronic)
SpringerBriefs in Applied Sciences and Technology
ISBN 978-3-319-22293-6   ISBN 978-3-319-22294-3   (eBook)
DOI 10.1007/978-3-319-22294-3

Library of Congress Control Number: 2015946994

Springer Cham Heidelberg New York Dordrecht London

Printed on acid-free paper

Springer International Publishing AG Switzerland is part of Springer Science+Business Media
(www.springer.com)

Preface

Over the last years, the indoor environmental quality of buildings has been extensively studied. The results obtained, by different researchers, have shown that indoor environmental conditions of buildings play an important role in terms of health and wellbeing of their occupants. The characterization of indoor hygro-thermal conditions is frequently pursued as part of the overall environmental evaluation. Different measurement techniques can be used and large amounts of data will become available after extensive in-situ campaigns. The application of adequate statistical tools and data mining techniques is therefore crucial to produce an adequate synthesis of the results to support sound conclusions.

The main benefit of the book is that it explores available methodologies for both conducting in situ measurements and adequately explore the results, based on a case study that illustrates the benefits and difficulties of concurrent methodologies.

The case study corresponds to a set of 25 similar houses where an extensive in situ measurement campaign was conducted. The dwellings are located in the same quarter, in Porto, Portugal. Measurements included indoor temperature and relative humidity, with continuous log in different rooms of each dwelling, blower-door tests, and complete outdoor conditions provided by a nearby weather station.

This book will include a variety of scientific and engineering disciplines, such as building physics, probability and statistics, and civil engineering. It is divided into several chapters that intend to be a synthesis of the current state of knowledge for the benefit of professional colleagues.

The authors acknowledge with gratitude the support received from the University of Porto—Faculty of Engineering, Portugal, namely the Laboratory of Building Physics (LFC). Finally, the authors would welcome reader comments, corrections, and suggestions with the aim of improving any future editions.

Nuno M.M. Ramos
João M.P.Q. Delgado
Ricardo M.S.F. Almeida
Maria L. Simões
Sofia Manuel

Acknowledgments

This work was developed on the framework of the project FCOMP-01-0124-FEDER-041748 and EXPL/ECM-COM/1999/2013 funded by FEDER funds through the Programa Operacional Factores de Competitividade—COMPETE and by National Funds through the FCT—Fundação para a Ciência e a Tecnologia. João M.P.Q. Delgado would like to thank Fundação para a Ciência e a Tecnologia for financial support through the grant SFRH/BPD/84377/2012.

Contents

Contents

Chapter 1
Introduction

Abstract This chapter explains the meaning of Data Mining and presents the motivation to apply it to indoor hygrothermal conditions evaluation since the data collection of connected parameters has increased in recent years due to measuring technology advances.

1.1 Motivation

Indoor hygrothermal conditions are frequently measured, in dwellings or public and commercial buildings, to support the assessment of indoor environmental quality. That assessment may be linked to different objectives, ranging from different storage facilities control to human quality of life evaluation. The devices applied on the measurements have evolved from thermo-hygrographs (see Fig. 1.1) that recorded temperature and relative humidity in paper to small dataloggers (see Fig. 1.2) with embedded sensors with a large storage capacity. A current variation of this type of systems is wireless transmission to a server, making data collection easy and less time consuming. The large quantities of indoor hygrothermal conditions data that have become available due to the evolution of the available monitoring systems creates the opportunity to have a deeper insight, providing the right analysis strategy is applied.

In this work, Data Mining was elected as the strategy to find that deeper insight. Data Mining is the popular term for what Hand et al. (2001) considers would be better named as "Knowledge mining for data". The same author includes it as part of a wider process of "knowledge discovery" that includes the following steps: data cleaning, data integration, data selection, data transformation, data mining, pattern evaluation and knowledge presentation. In this scheme, Data mining is the process where intelligent methods are applied to derive data patterns. Han and Kamber (2006) define Data Mining as the analysis of observational data sets to find unsuspected relationships and summarize the data in novel ways. The latter authors also explain that Data Mining is often used as secondary data analysis since the existing data was often collected for other purposes.

© The Author(s) 2016

N.M.M. Ramos et al., *Application of Data Mining Techniques in the Analysis of Indoor Hygrothermal Conditions*, SpringerBriefs in Applied Sciences and Technology, DOI 10.1007/978-3-319-22294-3_1

Fig. 1.1 Thermo-hygrograph

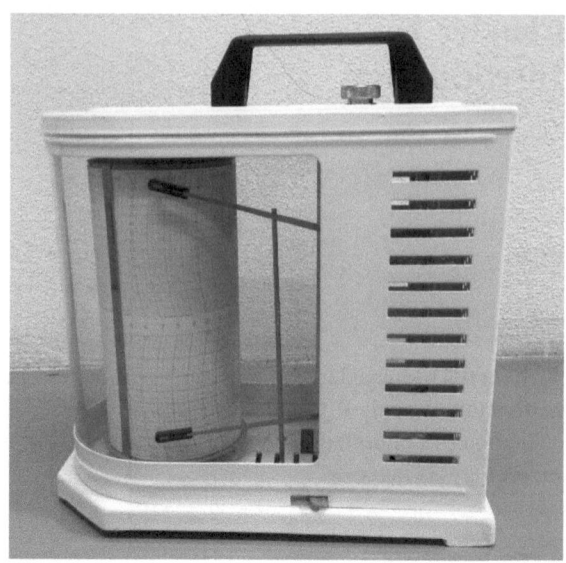

Fig. 1.2 Hygrothermal
datalogger

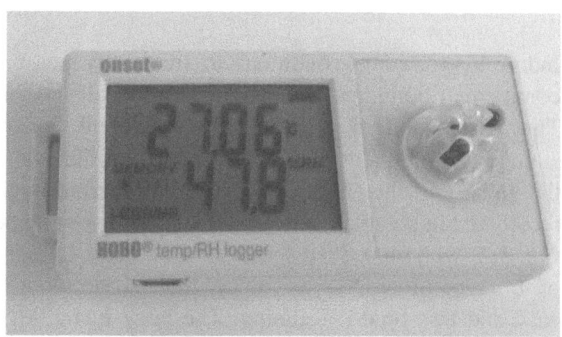

Hence, Data Mining application in the analysis of indoor hygrothermal conditions arises as an interesting topic as more and more sets of data are available, providing opportunities to find patterns and correlations. Those findings can then be applied to detect processes difficult to model, such as user behaviour.

1.2 Methodology

The methodology applied in this work comprises the following steps:

Chapter 2: The evaluation of indoor hygrothermal conditions is described, supported by a literature review. Involved parameters and standardized methodologies are summarized. The procedures for evaluation of human comfort are also briefly described.

Chapter 3: Data mining techniques applicable to the analysis of indoor hygro-thermal conditions are thoroughly described. Basic statistical tools are addressed first, followed by multivariate data techniques that allow for pattern analysis of the data.

Chapter 4: A case study is presented highlighting the difficulties associated with the analysis of available data sets.

Chapter 5: The data mining techniques described in Chap. 3 are applied to the case study described in Chap. 4. The overcome of the difficulties associated with the analysis process of the data using appropriate techniques is demonstrated.

Chapter 6: The main conclusions are presented.

References

Han, J., & Kamber, M. (2006). *Data mining: concepts and techniques*. Elsevier: Morgan Kaufman.

Hand, D., Mannila, H. & Smyth, P. (2001). *Principles of data mining*. The MIT Press.

Chapter 2
Indoor Hygrothermal Conditions

Abstract The evaluation of indoor hygrothermal conditions is described, supported by a literature review. Involved parameters and standardized methodologies are summarized. The procedures for evaluation of human comfort are also briefly described.

2.1 Involved Parameters

The search for a safe and comfortable environment has always been a major concern for humanity. In recent decades the occupancy levels of the buildings, the construction practices (lower air permeability of the envelope and the generalized use of heating, ventilation and air conditioning (HVAC) systems) and the users' expectations have dramatically changed, leading to a growing interest in the theme of the indoor environmental quality. In fact, nowadays the indoor environmental quality is an important factor for the health, comfort and performance of populations, since in developed areas of the planet people spend most of their time inside buildings (Wargocki 2009). The concept of indoor environmental quality is very broad and depends on many variables such as temperature, relative humidity, air velocity, air flow, occupancy, concentration of pollutants, noise, lighting... Yet, thermal comfort is unanimously recognised as crucial for an adequate indoor environmental quality (Alfano et al. 2010).

The classic definition of thermal comfort is the one presented by Fanger (1970) describing it as "the state of mind in which a person expresses satisfaction with the thermal environment". Afterwards several authors defended that satisfaction with the thermal environment depends, in addition to the physical factors that determine the heat exchange between the human body and the environment in which he is located (thermal balance), of other factors such as social, cultural and

© The Author(s) 2016

N.M.M. Ramos et al., *Application of Data Mining Techniques in the Analysis of Indoor Hygrothermal Conditions*, SpringerBriefs in Applied Sciences and Technology, DOI 10.1007/978-3-319-22294-3_2

psychological, which justify the different perceptions and responses for the same sensory stimuli. Therefore, users' past experiences and expectations may play a key role.

Several researchers had addressed their attention to the evaluation and quantification of thermal comfort in indoor environments. The main idea is to better understand which variables are involved, how it can be achieved, its impact in terms of occupants' health and productivity and how it can be quantified. Classical theories consider that thermal comfort depends on both individual and environmental factors as follows:

- Individual factors: metabolic rate, M [met]; and clothing insulation, I_{CL} [clo];
- Environmental factors: air temperature, T_a [°C]; mean radiant temperature, T_{mr} [°C], air velocity, v_{ar} [m/s]; and water vapor pressure, p_a [Pa].

For many years, the correct combination of these environmental factors which leads to comfort conditions has been pursued and several models to quantify an indoor environment on a single hygrothermal index have been proposed. Yet, in long term monitoring and in building simulation, comfort is commonly assessed just by the air temperature and relative humidity (Almeida and Freitas 2014; Olesen et al. 2011; Barbosa et al. 2015) and several simplified models, some included in national regulations and international standards, only use these two parameters, or simply the air temperature, to establish comfort conditions.

2.1.1 Air Temperature

Air temperature is the most important variable for thermal comfort quantification, since the sense of comfort in based on heat exchanges between body and environment and, therefore, enhanced by the temperature gradient between them (Lamberts 2005). Sometimes, only air temperature is used to establish comfort conditions. For instance, the Portuguese regulation defines thermal comfort based on air temperature: 20 °C in winter and 25 °C in summer.

2.1.2 Relative Humidity

Although often underestimated or even ignored compared to air temperature, relative humidity affects thermal comfort (ISO 2005; ASHRAE 2010), indoor air quality perception (Fang et al. 1998), occupants' health (Bornehag et al. 2001) and even building's energy consumption (Simonson 2000). Concerning thermal comfort, instead of global comfort, relative humidity is often linked to local comfort (Simonson et al. 2001).

2.2   Standardized Methodologies

In the 70s, based on Fanger's studies, ASHRAE presented a seven-level scale for thermal comfort assessment (Table 2.1). This scale became dominant in thermal comfort studies, being adopted in ISO 7730 (ISO 2005) and ASHRAE 55 (ASHRAE 2010) standards.

Fanger (1970) derived a general equation of comfort that attempts to include the effect of both individual and environmental factors. This index estimates the average vote for a group of persons of different nationalities, ages and sexes, according to the previous mentioned scale (Table 2.1) and was designated as *Predicted Mean Vote* (PMV). Fanger also suggested that the percentage of people who considered the environment as uncomfortable (feeling hot or cold) is related to their average vote, defining a second index called the *Predicted Percentage Dissatisfied* (PPD). PMV and PPD are commonly used as reference values in international standards to establish comfort conditions.

Yet, for long term monitoring of buildings' performance in service conditions, simpler methods based on air temperature and relative humidity are often used.

A simplified graphical method (Fig. 2.1) to evaluate thermal comfort is proposed in ASHRAE 55 (ASHRAE 2010). This method is applicable to environments with air velocity below 0.2 m/s, where the occupants' activities are sedentary (ranging between 1.0 met and 1.3 met) and clothing insulation varies from 0.5 to 1.0 clo. The comfort zone is for 80 % occupant acceptability, resulting from the combined effect of 10 % dissatisfied due to discomfort related to the whole body and 10 % that may occur from local thermal discomfort. The method establishes a comfort zone on a psychometric chart, requiring the operative temperature, which for moderate environments is often approximated by the air temperature, and the humidity ratio as inputs.

Adaptive models such as the ones suggested in ASHRAE 55 (2010) and EN 15271 (2007) only use air temperature as indicator of comfort.

ASHRAE 55 (2010) proposes a graphical method for indoor thermal comfort evaluation (Fig. 2.2), which can be applied to spaces where the occupants are engaged in near-sedentary physical activities, with metabolic rates ranging from 1.0 to 1.3 met. The base equation of the model was proposed by Brager et al. (2004), which establishes the indoor operative temperature, T_{oc}, as follows:

Table 2.1 Thermal comfort scale (adapted from ISO 2005 and ASHRAE 2010)

+3	Hot	Uncomfortable
+2	Warm	
+1	Slightly warm	Comfortable
0	Neutral	
−1	Slightly cool	
−2	Cool	Uncomfortable
−3	Cold	

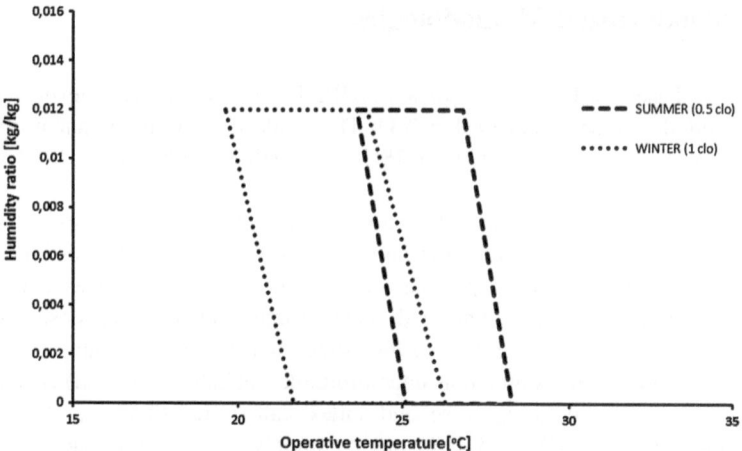

Fig. 2.1 Graphic comfort zone method (adapted from ASHRAE 2010)

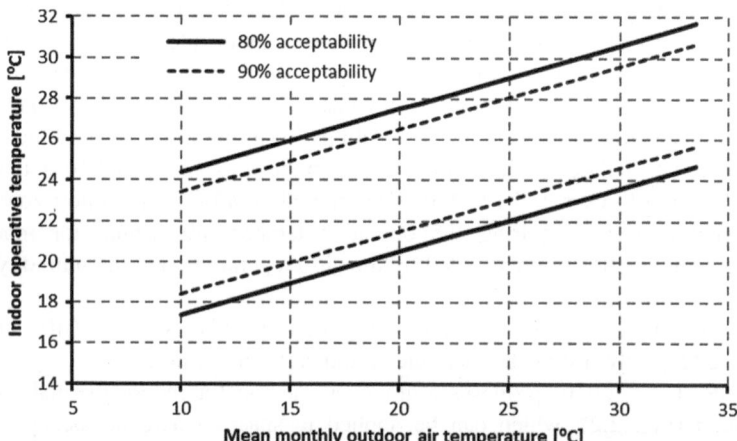

Fig. 2.2 ASHRAE adaptive model (adapted from ASHRAE 2010)

$$T_{oc} = 17.8 + 0.31 \cdot T_m \qquad\qquad (2.1)$$

in which

T_{oc} [°C] Indoor operative temperature
T_m [°C] Mean monthly outdoor air temperature

EN 15251 (2007) proposes another graphical method for thermal comfort evaluation (Fig. 2.3). This model defines the operative temperature as function of the weekly running mean outdoor temperature as follows:

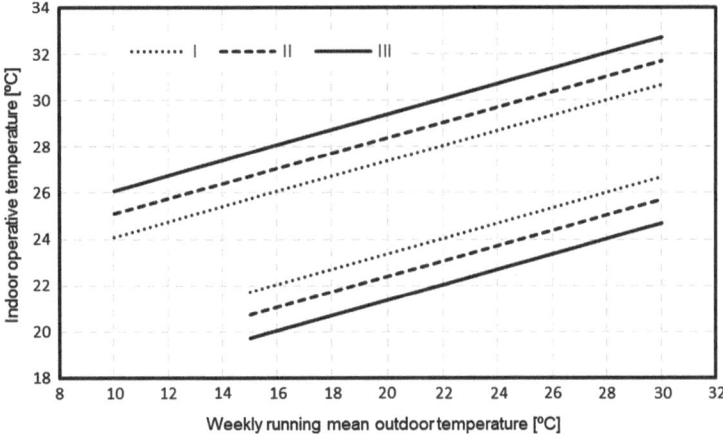

Fig. 2.3 EN 15251 adaptive model (EN 15251 2007)

$$T_{oc} = 0.33 \cdot T_{mp} + 18.8 \qquad (2.2)$$

in which

T_{mp} [°C] Weekly running mean outdoor temperature

This model is applicable in buildings without cooling devices and considers three categories, which establish the users' comfort demand (dissatisfied percentage limit requirement). Each category assumes a comfort temperature interval that corresponds to the distance between lower and upper limits in Fig. 2.2. The model is mainly suitable for summer conditions but can also be used for winter season (weekly running mean outdoor temperature between 10 and 15 °C) assuming the same temperature limits as for mechanically ventilated buildings. This method is valid for office buildings and other buildings of similar type used mainly for human occupancy with mainly sedentary activities and dwellings, where there is easy access to operable windows and occupants may freely adapt their clothing to the indoor and/or outdoor thermal conditions.

Although applicable for long term evaluation of the indoor environment, there are no normalized procedures establishing the required sample size and how the results should be analysed when one uses these models. Commonly, the interpretation of the results is performed with simple indicators such as the "total hours of discomfort". If additional information as buildings' characteristics, occupancy, energy consumption or other is also available, a large number of data must therefore be combined and data-mining techniques arise as valuable tools for finding meaningful correlations among them.

2.3   Literature Review

The ease of access to portable sensors for measuring and recording both the users' behaviour and the indoor environmental conditions of buildings, generalized conducting long term monitoring campaigns whose result is commonly a very large number of data whose interpretation is not always easy. On the other hand, the use of modern data mining techniques has evolved considerably since the 90s, starting its widespread use in fields as diverse as economics, sociology or computer science. However, the use of these tools in the area of building science only began to emerge in significant number in recent years.

In this sense, in recent years, some research where data mining techniques are applied in the construction area has emerged. These include, for example, studies which seek to relate the performance of buildings with the users' behaviour or seeking to define the indicators that best represent the building's performance.

Data mining techniques, by enabling to extract new and relevant information from a large number of data, are especially suitable for works of this nature, where repeatedly arise phenomena whose quantification is difficult, as in the case of the users' behaviour, clearly affected by both physiological and psychological parameters.

2.3.1   Building Performance Versus User Behaviour

The recognition of behavioural patterns from a wide range of observations is accepted as a high potential alternative. These patterns tend to anticipate and replicate actions usually repeated by users. These patterns are crossed with performance measurements of buildings, seeking to correlations between them. This approach has been used to explain aspects related to the energy efficiency and the hygrothermal performance of buildings.

D'Oca and Hong (2014) used cluster analysis and association rules to identify valid window operational patterns in measured data. Yang et al. (2015) studied the role of households' attitudes in building's energy consumption by analysing the behavioural variability. Ren et al. (2015) applied data mining techniques to understand the operation and performance of space heating systems for improving the occupant comfort while reducing energy use. Brown et al. (2015) investigated the impact of physical, behavioural and demographic variables in the buildings' energy and water consumption.

2.3.2   Building Performance Indicators

Post-occupancy assessment is crucial in the evaluation of building performance as a well-known gap exists between the predicted and real performance of buildings

(Meneses et al. 2012). There is a growing interest in this issue and post-occupancy evaluation assumes as the key for a better understanding on how the design process can be improved. Most of the approaches try to establish key performance indicators which characterize the buildings' consumption and performance patterns.

Lourenço et al. (2014) selected schools energy key performance indicators from the results of a survey conducted in eight Portuguese secondary schools. Shahrokni et al. (2014) evaluated the energy efficiency potential of the city of Stockholm from the billing meter data of the housing stock. Tian et al. (2014) used an office building as case study for applying bootstrap techniques to improve the accuracy of the model selected for thermal performance analysis. Goçer et al. (2015) discussed the importance of post-occupancy evaluation for an effective building design process.

References

Alfano, F. R. D. A., Bellia, L., Boerstra, A., Dijken, F. V., Ianniello, E., & Lopardo, G. et al. (2010). Indoor Environment and Energy Efficiency in Schools—Part 1 Principles REHVA—Federation of European Heating and Air-conditioning Associations, 2010.

Almeida, R. M. S. F., & Freitas, V. P. (2014). Indoor environmental quality of classrooms in Southern European climate. *Energy and Buildings, 81*, 127–140.

ASHRAE—American Society of Heating, Refrigerating and Airconditioning Engineers. (2010). Ansi/ASHRAE Standard 55-2010. Thermal Environmental Conditions for Human Occupancy. ASHRAE, Atlanta, USA.

Barbosa, R., Vicente, R., & Santos, R. (2015). Climate change and thermal comfort in Southern Europe housing: A case study from Lisbon. *Building and Environment, 92*, 440–451.

Borhehag, C. G., Blomquist, G., & Gyntelberg, F. (2001). Dampness in buildings and health—Nordic inter-disciplinary review of the scientific evidence on associations between exposure to "dampness" in buildings and health effects (NORDDAMP). *Indoor Air, 11*, 72–86.

Brager, G., Paliaga, G., & de Dear, R. (2004). The Effect of Personal Control and Thermal Variability on Comfort and Acceptability. ASHRAE—RP-1161—Final Report, ASHRAE, Atlanta, USA.

Brown, C., Gorgolewski, M., & Goodwill, A. (2015). Using physical, behavioral, and demographic variables to explain suite-level energy use in multiresidential buildings. *Building and Environment, 89*, 308–317.

D'Oca, S., & Hong, T. (2014). A data-mining approach to discover patterns of window opening and closing behavior in offices. *Building and Environment, 82*, 726–739.

EN 15251. (2007). EN 15251:2007—Indoor environmental input parameters for design and assessment of energy performance of buildings-addressing indoor air quality, thermal environment, lighting and acoustics. CEN—Comité Européen de Normalisation, Brussels, Belgium.

Fang, L., Clausen, G., & Fanger, P. (1998). Impact of temperature and humidity on the perception of indoor air quality. *Indoor Air, 8*, 80–90.

Fanger, P. O. (1970). *Thermal comfort. Analysis and applications in environmental engineering.* Denmark: McGraw-Hill.

Goçer, O., Hua, Y., & Goçer, K. (2015). Completing the missing link in building design process: Enhancing post-occupancy evaluation method for effective feedback for building performance. *Building and Environment, 89*, 14–27.

ISO—International Organization for Standardization. (2005). ISO 7730—Ergonomics of the Thermal Environment, Analytical Determination and Interpretation of Thermal Comfort using

Calculation of the PMV and PPD Indices and Local Thermal Comfort Criteria. ISO, Genève, Switzerland.

Lamberts, R, (2005). Desempenho térmico de edificações. Relatório do Laboratório de Eficiência Energética em Edificações, Universidade Federal de Santa Catarina, Florianópolis, Brasil.

Lourenço, P., Pinheiro, M. D., & Heitor, T. (2014). From indicators to strategies: Key Performance Strategies for sustainable energy use in Portuguese school buildings. *Energy and Buildings, 85*, 212–224.

Meneses, A. C., Cripps, A., Bouchlaghem, D., & Buswell, R. (2012). Predicted versus actual energy performance of non-domestic buildings: using post-occupancy evaluation data to reduce the performance gap. *Applied Energy, 97*, 355–364.

Olesen, B. W., Corgnati, S. P., & Raimondo, D. (2011). Evaluation methods for long term indoor environmental quality. *Proceedings of the Conference Indoor Air 2011*, June 5–10, Austin, Texas, USA.

Ren, X., Yan, D., & Hong, T. (2015). Data mining of space heating system performance in affordable housing. *Building and Environment, 89*, 1–13.

Shahrokni, H., Levihn, F., & Brandt, N. (2014). Big meter data analysis of the energy efficiency potential in Stockholm's building stock. *Energy and Buildings, 78*, 153–164.

Simonson, C. J. (2000). Moisture, thermal and ventilation performance of Tapanila ecological house. IN VTT (Ed.), VTT Research Notes, VTT, Espoo, Finland.

Simonson, C. J., Salonvaara, M., & Ojanen, T. (2001). *Improving indoor climate and confort with wooden structures*. Espoo, Finland: VTT Publications.

Tian, W., Song, J., Li, Z., & de Wilde, P. (2014). Bootstrap techniques for sensitivity analysis and model selection in building thermal performance analysis. *Applied Energy, 135*, 320–328.

Wargocki, P. (2009). Ventilation, thermal comfort, health and productivity. In D. Mumovic & M. Santamouris (Eds.), *A handbook of sustainable building design and engineering—an integrated approach to energy, health and operational performance*. London: Earthscan (Ch. 14).

Yang, S., Shipworth, M., & Huebner, G. (2015). His, hers or both's? The role of male and female's attitudes in explaining their home energy use behaviours. *Energy and Buildings, 96*, 140–148.

Chapter 3
Data Mining Techniques

Abstract This chapter presents available data mining techniques that can be of interest for application in indoor environment analysis. Descriptive statistics tools are presented with the focus on probability distribution and correlation analysis. Multivariate data techniques are also addressed, with a special focus on principal components determination and cluster analysis.

3.1 Principles

Data mining is the analysis of data for relationships that have not previously been discovered. Generally, data mining (sometimes called data or knowledge discovery) is the process of analyzing data from different perspectives and summarizing it into useful information—information that can be used to increase revenue, cuts costs, or both. Technically, data mining is the process of finding correlations or patterns among dozens of fields in large relational databases.

3.2 Basic Statistical Tools

Statistics is a field of mathematics that pertains to data analysis. Statistical methods and equations can be applied to a data set in order to analyze and interpret results, explain variations in the data, or predict future data. Statistics is important in the field of engineering as it provides tools to analyze collected data (Montgomery and Runger 2010).

© The Author(s) 2016

N.M.M. Ramos et al., *Application of Data Mining Techniques in the Analysis of Indoor Hygrothermal Conditions*, SpringerBriefs in Applied Sciences and Technology, DOI 10.1007/978-3-319-22294-3_3

3.2.1 Descriptive Statistics

Descriptive statistics can be useful for two purposes:

(1) to provide basic information about variables in a dataset and
(2) to highlight potential relationships between variables.

The most common descriptive statistics can be displayed graphically or pictorially and are measures of:

- Graphical/Pictorial methods
- Measures of central tendency
- Measures of dispersion
- Measures of association

The simplest way to measure the key characteristics of a data set is to estimate the summary statistics for the data. For a data series, $x_1, x_2, x_3, \ldots, x_n$, where n is the number of observations in the series, the most widely used summary statistics are as follows:

- The mean (μ), which is the average of all of the observations in the data series;
- The median, which is the midpoint of the series; half the data in the series is higher than the median and half is lower;
- The variance, which is a measure of the spread in the distribution around the mean and is calculated by first summing up the squared deviations from the mean, and then dividing by either the number of observations (if the data represent the entire population) or by this number, reduced by one (if the data represent a sample).
- The standard deviation is the square root of the variance.

The mean and the standard deviation are called the first two moments of any data distribution. A normal distribution can be entirely described by just these two moments; in other words, the mean and the standard deviation of a normal distribution suffice to characterize it completely. If a distribution is not symmetric, the skewness is the third moment that describes both the direction and the magnitude of the asymmetry and the kurtosis (the fourth moment) measures the thickness of the tails of the distribution relative to a normal distribution.

3.2.2 Probability Distributions

All probability distributions can be classified as discrete probability distributions or as continuous probability distributions, depending on whether they define probabilities associated with discrete variables or continuous variables. If a variable can take on any value between two specified values, it is called a continuous variable; otherwise, it is called a discrete variable. Many probability distributions are commonly used in the field of engineering (Haldar and Mahadevan 2000).

3.2.2.1 Discrete Probability Distributions

- **Bernoulli**
  The *Bernoulli* distribution is a discrete distribution having two possible out-comes labelled by $x = 0$ and $x = 1$ in which $x = 1$ ("success") occurs with probability p and $x = 0$ ("failure") occurs with probability $q = 1 - p$, where $0 < p < 1$. It therefore has probability mass function

$$f(x) = p^x \cdot (1 - p)^{1-x} = \begin{cases} 1 - p & \text{for } x = 0 \\ p & \text{for } x = 1 \end{cases}. \tag{3.1}$$

  The mean of the distribution (μ) is equal to p and the variance (σ^2) is equal to $p \cdot (1 - p)$.

- **Binomial**
  A binomial random variable is the number of successes x in n repeated *bernoulli* trials (where the result of each *bernoulli* trial is true with probability p and false with probability $q = 1 - p$). The binomial probability mass function is

$$f(x) = \binom{n}{x} p^x \cdot (1 - p)^{n-x} \quad \text{for } x = 0, 1, 2, \ldots, n \tag{3.2}$$

  The mean of the distribution (μ) is equal to $n.p$ and the variance (σ^2) is equal to $n \cdot p \cdot (1 - p)$.

- **Poisson**
  The Poisson distribution is used to model the number of events occurring within a given time interval. The formula for the *Poisson* probability mass function is

$$f(x) = e^{-\lambda} \frac{\lambda^x}{x!} \quad \text{for} \quad x = 0, 1, 2, \ldots \tag{3.3}$$

  where λ is the shape parameter which indicates the average number of events in the given time interval.
  The mean of the distribution (μ) is equal to λ and the variance (σ^2) is also equal to λ.

3.2.2.2 Continuous Probability Distributions

- **Normal or Gaussian**
  The *Normal* or *Gaussian* distribution is a very important statistical distributions. The *Normal* distributions is symmetric and have bell-shaped density curves with a single peak. A *Normal* or *Gaussian* distribution in a variable X with mean μ and variance σ^2 is a statistic distribution with probability density function

$$f(x) = \frac{1}{\sigma\sqrt{2\pi}} e^{-\frac{(x-\mu)^2}{2\sigma^2}} \quad \text{for } x \in (-\infty, +\infty) \tag{3.4}$$

The case where $\mu = 0$ and $\sigma = 1$ is called the standard normal distribution.
A *Lognormal* distribution is a continuous distribution in which the logarithm of
a variable has a normal distribution. A *Lognormal* distribution results if the
variable is the product of a large number of independent, identically-distributed
variables in the same way that a normal distribution results if the variable is the
sum of a large number of independent, identically-distributed variables.

- **Exponential**
  Given a *Poisson* distribution with rate λ, the distribution of waiting times
  between successive events is

$$F(x) = P(X \le x) = 1 - e^{-\lambda x}, \tag{3.5}$$

and the probability density function is

$$f(x) = \lambda \cdot e^{-\lambda x} \quad \text{for } x \ge 0. \tag{3.6}$$

The mean of the distribution (μ) is equal to $1/\lambda$ and the variance (σ^2) is also
equal to $1/\lambda^2$.

3.2.3 Correlation Matrices

Looking for relationships in the data, when there are two series of data, there are a
number of statistical measures that can be used to capture how the series move
together over time. The two most widely used measures of how two variables move
together (or do not) are the correlation and the covariance. For two data series, $X(x_1,
x_2, \ldots)$ and $Y(y_1, y_2, \ldots)$, the covariance provides a measure of the degree to which
they move together and is estimated by taking the product of the deviations from
the mean for each variable in each period:

$$\text{cov}(X, Y) = E[(X - \mu_X)(Y - \mu_Y)], \tag{3.7}$$

where E is the expected value operator, μ_X and μ_Y are the expected values of
variables X and Y, respectively.

The sign on the covariance indicates the type of relationship the two variables
have. A positive sign indicates that they move together and a negative sign that they
move in opposite directions. Although the covariance increases with the strength of
the relationship, it is still relatively difficult to draw judgments on the strength of the
relationship between two variables by looking at the covariance, because it is not
standardized.

The correlation coefficient ρ_{XY} is the standardized measure of the relationship between two random variables. It can be computed from the covariance:

$$\rho_{XY} = \frac{\text{cov}(X, Y)}{\sigma_X \cdot \sigma_Y}, \tag{3.8}$$

where σ_X and σ_Y are the expected values of variables X and Y, respectively.

The correlation coefficient can never be greater than one or less than negative one. A correlation close to zero indicates that the two variables are unrelated. A positive correlation indicates that the two variables move together, and the relationship is stronger as the correlation gets closer to one. A negative correlation indicates the two variables move in opposite directions, and that relationship gets stronger the as the correlation gets closer to negative one. Two variables that are perfectly positively correlated ($\rho_{XY} = 1$) essentially move in perfect proportion in the same direction, whereas two variables that are perfectly negatively correlated move in perfect proportion in opposite directions.

3.2.4 Multi-Way Frequency Tables

To present the construction of a multidimensional frequency table, it is necessary to introduce the basic terms of frequency tables for the two variables.

Let us assume that the number of categories in variable X is r and in variable Y is c, then the n_{ij} is the observed frequency of category i of variable $X(i = 1,\ldots, r)$ and of category j of variable $Y(j = 1,\ldots, c)$. For Pearson χ^2 test of independence should be determined successively:

- row sums (row frequencies) $n_{i\bullet} = \sum_{j=1}^{c} n_{ij}$;

- column sums (row frequencies) $n_{\bullet j} = \sum_{i=1}^{r} n_{ij}$.

Row and column sums give information about the total count of categories of both variables. Next there are observed proportion $p_{ij} = \frac{n_{ij}}{n}$ which is the percentage share of occurrence in the study of category i of variable X and of category j of variable Y. These values are elements of the matrix P. On this basis we shall determine the row proportion:

$$p_{i\bullet} = \sum_{j=1}^{c} p_{ij} \tag{3.9}$$

and column proportion:

$$p_{\bullet j} = \sum_{i=1}^{r} p_{ij}. \tag{3.10}$$

These values show the percentage of occurrence of the selected category in the grand total. The row proportions are denoted as vector r, column proportions as vector c. Finally, there is expression of the expected proportions, $\hat{p}_{ij} = p_{i\bullet} \cdot p_{\bullet j}$, and expected frequencies, $n \cdot \hat{p}_{ij}$.

The χ^2 test of independence or likelihood ratio test is used for evaluating variables dependence. If all variables are equally important for the research problem, it is possible to create many different multi-way tables. One or both of the independence tests should be done for each of these tables in the following way.

When variables X and Y are independent $p_{ij} = p_{i\bullet} \cdot p_{\bullet j}$. If the hypothesis of independence is true, the χ^2 statistics

$$\chi^2 = \sum_{i=1}^{r} \sum_{j=1}^{c} \frac{\left(n_{ij} - n \cdot p_{i\bullet} \cdot p_{\bullet j}\right)^2}{n \cdot p_{i\bullet} \cdot p_{\bullet j}} \tag{3.11}$$

has a χ^2 distribution with $(r-1)(c-1)$ degrees of freedom.

In the Likelihood ratio approach the statistics

$$L^2 = 2 \cdot \sum_{i=1}^{r} \sum_{j=1}^{c} n_{ij} \ln\left(\frac{n_{ij}}{\hat{n}_{ij}}\right), \tag{3.12}$$

where $\hat{n}_{ij}$ are expected frequencies; L^2 has a χ^2 distribution with $(r-1)(c-1)$ degrees of freedom if the hypothesis of independence is held.

3.3   Multivariate Data Techniques

3.3.1   Principal Components and Factor Analysis

In many studies the number of variables is considered too large to be treated, making it often absolutely necessary to reduce the scale of analysis for the situation becomes understandable, that is, it becomes necessary to use a reduction of data technique (Han and Kamber 2006).

Factorial analysis (or analysis of common factors) and principal component analysis are statistical techniques which aim to represent or describe an initial number of variables by a smaller number of hypothetical variables (factors/main components). That is, it allows to identify new variables (factors/main components), less than the original set, but without significant loss of information contained in this set.

The general purpose of such techniques is to find a way to condense (summarize) information contained in the original set of variables, into a smaller set of variables losing the least amount of information. These are, therefore, data reduction techniques used to investigate the relationships (correlations) between the variables and describe them, if possible, in terms of a smaller number of variables, called factors/main components.

3.3.1.1 Principal Components

Often it is intended to describe a data set consisting of n individuals characterized by p quantitative variables. This kind of data leads to an initial dissymmetric table Q whose general term q_{ij} is the value taken by the jth variable in the individual i. The variables can have different units of measure with very different mean values, suggesting that the data must be centered. The initial table Q becomes a table X, whose general term is

$$x_{ij} = \frac{1}{\sqrt{n}} \left(q_{ij} - \bar{q}_j \right), \tag{3.13}$$

where $\bar{q}_j$ is the arithmetic mean of the values taken by the variable j. The matrix to diagonalize $X^T X$ is the variance-covariance matrix.

Often an additional modification of the initial table is still required, when the dispersion of the variables is very different or when the variables differ in their nature and are expressed in units of measurement not comparable. This problem can be addressed by standardizing variables, that is, making them dimensional with zero mean and unit variance. The general term of table X, in this case, is given by

$$x_{ij} = \frac{1}{\sqrt{n}} \frac{q_{ij} - \bar{q}_j}{s_j}, \tag{3.14}$$

where s_j is the standard deviation of variable j. The matrix $X^T X$ becomes the matrix of experimental correlations.

The transformation given by Eq. (3.13) causes a cloud translation, to align the centre of gravity with the origin. The influence the overall level of each variable is thus eliminated. The quotient $\frac{1}{\sqrt{n}}$ it is intended to match matrix $X^T X$ with the variance-covariance matrix.

The quotient by the standard deviation s_j causes the reduction of the effect of widely dispersed variables on the distances between individuals:

$$d^2(i, i') = \sum_{j=1}^{p} \left(x_{ij} - x_{i'j} \right)^2 = \frac{1}{n} \sum_{j=1}^{p} \left(\frac{q_{ij} - \bar{q}_j}{s_j} \right)^2. \tag{3.15}$$

Thus each variable will have a similar contribution in determining distances. In resume, the analysis of the cloud of points in R^p leads to translation from the source to the centre of gravity and the transformation of the scales of different axes. The analysis of the transformed picture is reflected in the survey of eigenvectors u_j of the experimental correlation matrix $R = X^T X$. The coordinates of the individuals in the factorial axes are given the following scalar products: $W = X\,u$.

The division by $s_j \sqrt{n}$, where R^p translated into a change of scale of the axes leads to deformation in a cloud R^n; each variable happens to be positioned at the unit distance from the origin.

$$d^2(j,0) = \frac{1}{n} \sum_{j=1}^{n} \left(\frac{q_{ij} - \bar{q}_j}{s_j} \right)^2 = 1. \tag{3.16}$$

The variables are positioned over a hypersphere with radius 1 centred at the origin. The distance between two points j e j' is given by:

$$d^2(j,j') = \frac{1}{n} \sum_{j=1}^{n} \left(\frac{q_{ij} - \bar{q}_j}{s_j} - \frac{q_{ij'} - \bar{q}_{j'}}{s_{j'}} \right)^2$$

$$d^2(j,j') = \frac{1}{n} \sum_{j=1}^{n} \left(\frac{q_{ij} - \bar{q}_j}{s_j} \right)^2 + \frac{1}{n} \sum_{j=1}^{n} \left(\frac{q_{ij'} - \bar{q}_{j'}}{s_{j'}} \right)^2 - 2 \frac{1}{n} \sum_{j=1}^{n} \left(\frac{q_{ij} - \bar{q}_j}{s_j} - \frac{q_{ij'} - \bar{q}_{j'}}{s_{j'}} \right)$$

$$d^2(j,j') = 2 \left(1 - r_{ij'} \right)$$

$$\tag{3.17}$$

which $r_{jj'}$ is the linear correlation coefficient between the variables j and j'. Thus, the proximity between the variables can be interpreted in terms of their correlations: the points are close if the correlation is strongly positive ($r_{jj'} \approx 1$) and faraway if it is strongly negative ($r_{jj'} \approx -1$). Intermediate distances are the independent variables ($r_{jj'} = 0$).

The variables coordinates in the factorial axes are given by

$$F = X^T v. \tag{3.18}$$

The variable coordinates in an axis are the correlation coefficients of the variables with the axis. Indeed, the coordinate $f_{j\alpha}$ of a variable j in the axes α, is given by

$$f_{j\alpha} = \sum_{i=1}^{n} x_{ij} \cdot v_{i\alpha} \tag{3.19}$$

Given the processing used and by the fact of the construction of the vectors v are with mean zero and unitary variance, the variable j in the coordinate axis α is given by

$$f_{j\alpha} = \sum_{i=1}^{n} \frac{\left(q_{ij} - \bar{q}_j\right)\left(q_{ij} - \bar{q}_j'\right)}{s_j} v_{i\alpha} = r_{j\alpha} \qquad (3.20)$$

which $r_{j\alpha}$ is the correlation coefficient between the variable j and the principal component α.

- **Algorithm of principal components analysis**
  The analysis algorithm in principal components normative can be described in different steps:

  Step 1 Transformation of the original data matrix. The Table Q is transformed in another matrix X by the reducing operation of the initial variables:

  $$x_{ij} = \frac{1}{\sqrt{n}} \frac{q_{ij} - \bar{q}_j}{s_j} \qquad (3.21)$$

  Step 2 Calculating the correlation matrix R, whose generic element is given by:

  $$r_{ij'} = \sum_{i=1}^{n} x_{ij} \cdot x_{ij'} = \frac{1}{n} \sum_{i=1}^{n} \frac{\left(q_{ij} - \bar{q}_j\right)\left(q_{ij'} - \bar{q}_{j'}\right)}{s_j \cdot s_{j'}} \qquad (3.22)$$

  Step 3 Diagonalization of the correlation matrix resulting p eigenvalues λ_α and p eigenvectors u_α.

  Step 4 Calculating the coordinates of the variables in the factorial axes, given by:

  $$f_{j\alpha} = \sum_{i=1}^{n} x_{ij} \cdot v_{i\alpha} \qquad (3.23)$$

  Step 5 Calculating the projections of individuals in factorial axes, given by:

  $$w_{i\alpha} = \sum_{j=1}^{p} x_{ij} \cdot u_{j\alpha} \qquad (3.24)$$

  Step 6 Selecting the size of the sub-space, whose accumulated inertia explain a sufficient percentage of the total inertia, according to criteria that are analysed below.

  Step 7 Projection of eventual individuals and in supplementary variables.

  Step 8 Interpretation of results.

- **Interpretation rules**

  The factorial axes resulting from a Principal Component Analysis constitute a new hierarchical basis space engendered by the data, whose total inertia is given by:

$$I_g = tr\left(X^T X\right) = \sum_{\alpha=1}^{p} \lambda_\alpha \qquad\qquad (3.25)$$

Each axis is responsible for a certain percentage of the cloud of inertia, given by:

$$100 \cdot \frac{\lambda_\alpha}{I_g} \qquad\qquad (3.26)$$

The basic aim of Principal Component Analysis is to reduce the size of the spaces at stake. A convenient way to visualize the cloud will project it in the plans defined by the factorial axes that represent together a percentage deemed sufficient inertia. There are several criteria to find the p_r number of axes to retain, shaking reducing the size of the space with the need to explain a significant proportion of the total variance.

We present below the most commonly criteria used (alone or combined):

1. A spherical cloud, without preferential stretching, the eigenvalues resulting from a normative in analysis are all the same: $\lambda_\alpha = 1, \quad \forall \alpha \in \{1, \ldots, p\}$.
   Then, one can choose p_r as number of axes α, such that $\lambda_\alpha \geq 1$.
2. If τ is a percentage of the total inertia fixed in advance, usually 80 %. Then p_r is the number of axes, such that $100 \sum_{g=1}^{p_r} \frac{\lambda_g}{I_g} \geq \tau$.
3. Considering the curve (*scree plot*) that lists the serial number of each axis with the actual value associated with it. If this curve shows a stabilization of eigenvalues, one can retain only the axes order with higher numbers that starts the stabilization (see Fig. 3.1).

It may happen that there are poorly explained variables in retained axes that have high correlations with unselected axes, using the above criteria. In this case it is advisable to also retain these axes. The overlap of the projections of the two clouds in the same plane becomes more significant interpretation, since it take some

Fig. 3.1 Distribution of eigen values

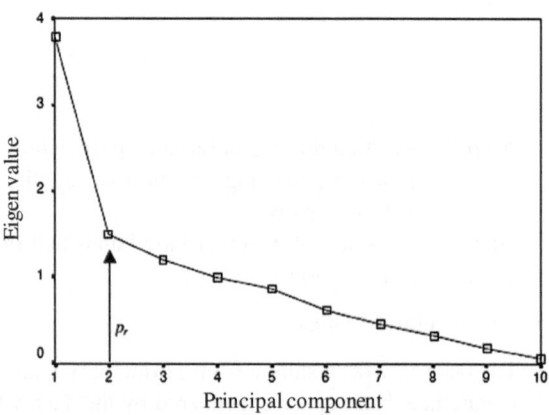

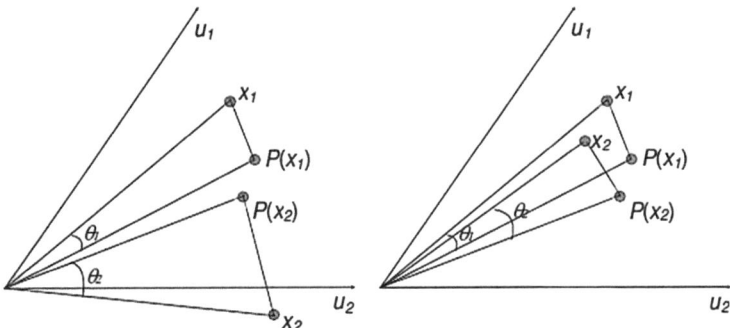

Fig. 3.2 Orthogonal projection on the plane $u_1 x\ u_2$

precautions. The clouds have different meanings for the interpretation of variables and individuals must be carried out independently. The close proximity between an individual and a variable does not have a very precise mathematical meaning. However the interpretation of factorial axes, based on correlations presenting with the variables, allow relating the two clouds in a roundabout way.

Before analysing the relative position of individuals or variables is necessary to determine their quality of representation in the relevant plan. The proximity of the projections do not necessarily correspond to a real proximity (Fig. 3.2).

In case I, the projections are close although individuals x_1 and x_2 are widely separated. The angles q_1 and q_2 are large. In case II, the angles are small, individuals x_1 and x_2 are close to their projections, and therefore are close together. The cosine of the angle formed by the vector x_i giving the individual's position with the plan is considered a good measure of that individual representation quality.

The representative points of the variables are in the hypersphere of radius 1. The representation quality of a variable can be evaluated directly tracing the unit circle: the variables positioned next the plan are projected along the circumference.

The inner product value of the vectors joining two cloud points in R^n is the correlation coefficient between the corresponding variables (also the cosine of the angle between the two vectors). Also, as mentioned above, the variables coordinates in an axis are the correlation coefficients of the variables with the axis.

Thus, analysis of nearby or oppositions between variables is done in have we correlations. In the example of Fig. 3.3 the projections of the 5 variables in the plan $u_1\ u_2$ are represented as the correlation circle.

The variables x_1, x_2, x_4 and x_5 are well represented in plan, as are close to the unit circle: x_1 and x_2 are strongly correlated, but are independent of the variables x_4 and x_5, which between themselves have a correlation strong negative. As for the variable x_3, misrepresented this plan, nothing can be concluded.

If the coordinates of the variables are interpretable in terms of correlations, the same does not happen with individuals. The analysis of cloud Rp is done in relation to the centre of gravity, and the Euclidean distance measure that quantifies the relationship (near and oppositions) between the points.

Fig. 3.3 Correlation circle

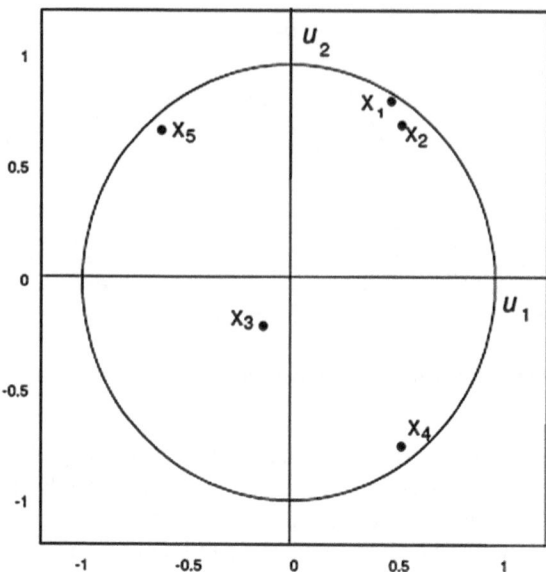

- **Individuals and additional variables**

  It often happens that know the values of variables p in a number of new individuals. It may be interesting to position these new individuals in the cloud already analysed. In other cases may be of interest to examine how additional points the centres of gravity of individuals belonging to the same category. It may also happen that new variables are measured on all the individuals or that has voluntarily been "set aside" because they wanted to keep only a group of homogeneous characteristics. In any of the above cases, the interpretation of the factors can be enriched projecting these variables in the illustrative cloud principal planes of the active variables. It can happen, therefore, that there are new rows and columns bordering the new data frame.

3.3.1.2   Factor Analysis

Factor Analysis is primarily used for data reduction or structure detection method. The purpose of data reduction is to remove redundant (highly correlated) variables from the data file, perhaps replacing the entire data file with a smaller number of uncorrelated variables. The purpose of structure detection is to examine the underlying (or latent) relationships between the variables. Therefore, the main applications of factor analytic techniques are:

(1) to *reduce* the number of variables and
(2) to *detect structure* in the relationships between variables, that is to *classify variables*.

3.3.2 Discriminant and Cluster Analysis

3.3.2.1 Discriminant Analysis

Given a set of independent variables, discriminant analysis attempts to find linear combinations of those variables that best separate the groups of cases. These combinations are called discriminant functions and have the form displayed in the equation:

$$d_{ik} = b_{0k} + b_{1k}x_{i1} + \cdots + b_{pk}x_{ip} \qquad (3.27)$$

where

d_{ik} is the value of the kth discriminant function for the ith case
p is the number of predictors
b_{jk} is the value of the jth coefficient of the kth function
x_{ij} is the value of the ith case of the jth predictor

The number of functions equals *min* (#groups-1, #predictors).

The procedure automatically chooses a first function that will separate the groups as much as possible. It then chooses a second function that is both uncorrelated with the first function and provides as much further separation as possible. The procedure continues adding functions in this way until reaching the maximum number of functions as determined by the number of predictors and categories in the dependent variable.

3.3.2.2 Cluster Analysis

The Cluster Analysis procedure is an exploratory tool designed to reveal natural groupings (or clusters) within a data set that would otherwise not be apparent. The algorithm employed by this procedure has several desirable features that differentiate it from traditional clustering techniques:

- The ability to create clusters based on both categorical and continuous variables.
- Automatic selection of the number of clusters.
- The ability to analyze large data files efficiently.

In order to handle categorical and continuous variables, the Cluster Analysis procedure uses a likelihood distance measure which assumes that variables in the cluster model are independent. Further, each continuous variable is assumed to have a normal (Gaussian) distribution and each categorical variable is assumed to have a multinomial distribution. Empirical internal testing indicates that the procedure is fairly robust to violations of both the assumption of independence and the distributional assumptions, but one should try to be aware of how well these assumptions are met.

The two steps of the Cluster Analysis procedure's algorithm can be summarized as follows:

Step 1   The procedure begins with the construction of a Cluster Features (CF) Tree. The tree begins by placing the first case at the root of the tree in a leaf node that contains variable information about that case. Each successive case is then added to an existing node or forms a new node, based upon its similarity to existing nodes and using the distance measure as the similarity criterion. A node that contains multiple cases contains a summary of variable information about those cases. Thus, the CF tree provides a capsule summary of the data file.

Step 2   The leaf nodes of the CF tree are then grouped using an agglomerative clustering algorithm. The agglomerative clustering can be used to produce a range of solutions. To determine which number of clusters is "best", each of these cluster solutions is compared using *Schwarz's Bayesian Criterion* (BIC) or the *Akaike Information Criterion* (AIC) as the clustering criterion.

- **Approaches to cluster analysis**

There are a number of different methods that can be used to carry out a cluster analysis; these methods can be classified as follows:

Hierarchical methods

- Agglomerative methods, in which subjects start in their own separate cluster. The two 'closest' (most similar) clusters are then combined and this is done repeatedly until all subjects are in one cluster. At the end, the optimum number of clusters is then chosen out of all cluster solutions.
- Divisive methods, in which all subjects start in the same cluster and the above strategy is applied in reverse until every subject is in a separate cluster. Agglomerative methods are used more often than divisive methods, so this handout will concentrate on the former rather than the latter.

Non-hierarchical methods (often known as k-means clustering methods)

- Types of data and measures of distance

The data used in cluster analysis can be interval, ordinal or categorical. However, having a mixture of different types of variable will make the analysis more complicated. This is because in cluster analysis you need to have some way of measuring the distance between observations and the type of measure used will depend on what type of data you have. A number of different measures have been proposed to measure 'distance' for binary and categorical data. For interval data the most common distance measure used is the Euclidean distance.

- Euclidean distance

In general, if there are p variables $X_1, X_2, ..., X_p$ measured on a sample of n subjects, the observed data for subject i can be denoted by $x_{i1}, x_{i2}, ..., x_{ip}$ and the observed data for subject j by x_{j1}, $x_{j2}, ..., x_{jp}$. The Euclidean distance between these two subjects is given by

$$d_{ij} = \sqrt{(x_{i1} - x_{j1})^2 + (x_{i2} - x_{j2})^2 + \cdots + (x_{ip} - x_{jp})^2} \qquad (3.28)$$

When using a measure such as the Euclidean distance, the scale of measurement of the variables under consideration is an issue, as changing the scale will obviously effect the distance between subjects (e.g. a difference of 10 cm could being a difference of 100 mm). In addition, if one variable has a much wider range than others then this variable will tend to dominate. For example, if body measurements had been taken for a number of different people, the range (in mm) of heights would be much wider than the range in wrist circumference, say. To get around this problem each variable can be standardized (converted to z-scores). However, this in itself presents a problem as it tends to reduce the variability (distance) between clusters. This happens because if a particular variable separates observations well then, by definition, it will have a large variance (as the between cluster variability will be high). If this variable is standardized then the separation between clusters will become less. Despite this problem, many textbooks do recommend standardization. If in doubt, one strategy would be to carry out the cluster analysis twice—once without standardizing and once with—to see how much difference, if any, this makes to the resulting clusters.

- Hierarchical agglomerative methods

Within this approach to cluster analysis there are a number of different methods used to determine which clusters should be joined at each stage. The main methods are summarized below.

Nearest neighbor method (single linkage method)

In this method the distance between two clusters is defined to be the distance between the two closest members, or neighbors. This method is relatively simple but is often criticized because it doesn't take account of cluster structure and can result in a problem called chaining whereby clusters end up being long and straggly. However, it is better than the other methods when the natural clusters are not spherical or elliptical in shape.

Furthest neighbour method (complete linkage method)

In this case the distance between two clusters is defined to be the maximum distance between members—i.e. the distance between the two subjects that are furthest apart. This method tends to produce compact clusters of similar size but, as for the nearest neighbour method, does not take account of cluster structure. It is also quite sensitive to outliers.

Average (between groups) *linkage method*

The distance between two clusters is calculated as the average distance between all pairs of subjects in the two clusters. This is considered to be a fairly robust method.

Centroid method

Here the centroid (mean value for each variable) of each cluster is calculated and the distance between centroids is used. Clusters whose centroids are closest together are merged. This method is also fairly robust.

Ward's method

In this method all possible pairs of clusters are combined and the sum of the squared distances within each cluster is calculated. This is then summed over all clusters. The combination that gives the lowest sum of squares is chosen. This method tends to produce clusters of approximately equal size, which is not always desirable. It is also quite sensitive to outliers. Despite this, it is one of the most popular methods, along with the average linkage method.

It is generally a good idea to try two or three of the above methods. If the methods agree reasonably well then the results will be that much more believable.

- Selecting the optimum number of clusters

As stated above, once the cluster analysis has been carried out it is then necessary to select the 'best' cluster solution. There are a number of ways in which this can be done, some rather informal and subjective, and some more formal. The more formal methods will not be discussed in this handout. Below, one of the informal methods is briefly described. When carrying out a hierarchical cluster analysis, the process can be represented on a diagram known as a dendrogram. This diagram illustrates which clusters have been joined at each stage of the analysis and the distance between clusters at the time of joining. If there is a large jump in the distance between clusters from one stage to another then this suggests that at one stage clusters that are relatively close together were joined whereas, at the following stage, the clusters that were joined were relatively far apart. This implies that the optimum number of clusters may be the number present just before that large jump in distance. This is easier to understand by actually looking at a dendrogram—see references for further information.

Non-hierarchical or k-means clustering methods

In these methods the desired number of clusters is specified in advance and the 'best' solution is chosen. The steps in such a method are as follows:

1. Choose initial cluster centers (essentially this is a set of observations that are far apart—each subject forms a cluster of one and its centre is the value of the variables for that subject).

2. Assign each subject to its 'nearest' cluster, defined in terms of the distance to the centroid.
3. Find the centroids of the clusters that have been formed
4. Re-calculate the distance from each subject to each centroid and move observations that are not in the cluster that they are closest to.
5. Continue until the centroids remain relatively stable.

Non-hierarchical cluster analysis tends to be used when large data sets are involved. It is sometimes preferred because it allows subjects to move from one cluster to another (this is not possible in hierarchical cluster analysis where a subject, once assigned, cannot move to a different cluster). Two disadvantages of non-hierarchical cluster analysis are:

(1) it is often difficult to know how many clusters you are likely to have and therefore the analysis may have to be repeated several times and
(2) it can be very sensitive to the choice of initial cluster centres. Again, it may be worth trying different ones to see what impact this has.

One possible strategy to adopt is to use a hierarchical approach initially to determine how many clusters there are in the data and then to use the cluster analysis.

3.3.3 Multiple Regression Analysis

Multiple Regression is a statistical tool that allows you to examine how multiple independent variables are related to a dependent variable. Once we have identified how these multiple variables relate to your dependent variable, we can take information about all of the independent variables and use it to make much more powerful and accurate predictions about why things are the way they are.

A multiple regression equation for predicting Y can be expressed a follows:

$$Y = a_0 + a_1 X_1 + a_2 X_2 + \cdots + a_n X_n \tag{3.29}$$

To apply the equation, each X_j score for an individual case is multiplied by the corresponding B_j value, the products are added together, and the constant A is added to the sum. The result is Y', the predicted Y value for the case.

For a given set of data, the values for B_js are determined mathematically to minimize the sum of squared deviations between predicted Y and the actual Y scores. Calculations are quite complex, and best performed with the help of a computer, although simple cases with only one or two predictors can be solved by hand with special formulas.

For the statistical test to be accurate, a set of assumptions must be satisfied. The key assumptions are that cases are sampled randomly and independently from the population, and that the deviations of Y values from the predicted Y values are normally distributed with equal variance for all predicted values of Y.

3.3.4   Classification Trees

Classification trees are used to predict membership of cases or objects in the classes of a categorical dependent variable from their measurements on one or more predictor variables. Classification tree analysis is one of the main techniques used in Data Mining. The goal of classification trees is to predict or explain responses on a categorical dependent variable, and as such, the available techniques have much in common with the techniques used in the more traditional methods of Discriminant Analysis, Cluster Analysis, Nonparametric Statistics, and Nonlinear Estimation.

References

Han, J., & Kamber, M. (2006). Data mining: Concepts and techniques. San Francisco: Morgan Kaufman, Elsevier.

Haldar, A., & Mahadevan, S. (2000). Probability, reliability and statistical methods in engineering design. New York: Wiley.

Montgomery, D., & Runger, G. (2010). Applied statistics and probability for engineers. New York: Wiley.

Chapter 4
Case Study

Abstract This chapter describes the case study, including sample description and presentation of the in situ measurements. The measurement results for both outdoor and indoor conditions are presented, highlighting the difficulty of deep analysis without using data mining techniques.

4.1 Sample Description

For the purpose of the study a neighbourhood located in Portugal was chosen. All the buildings have similar features, usually 4 storey high and constituted by two flats per floor connected to a common staircase that was originally enclosed. As for its structure all the buildings have a reinforced concrete frame with concrete slabs including top floor ceiling, the envelope consists of masonry cavity walls and ground floor units are slab on grade. The layout of the buildings can be seen in Fig. 4.1. The experimental campaign was carried out over a sample of 24 flats located at different story heights. The typologies ranged from T2 to T4 with corresponding net floor area ranging from 49 to 75 m^2. The occupation rates in the different typologies were relatively low. In some units, the effort made by the users to increase the comfort levels is easily noticed due to the modifications made in the laundry. The mentioned units belong to the older buildings group where the original solution for this room presented permanent openings. Another action common to most of the residents was to caulk the windows. The envelope modifications intended airtightness increase but were frequently combined with ventilation upgrading as well.

4.2 In-Situ Campaign

In-situ measurements were conducted in all the flats of the sample. The air tightness was measured with a Blower Door tests were performed using the Retrotec 1000 blower door model, according to standard EN 13829 (CEN 2000) method A,

© The Author(s) 2016
N.M.M. Ramos et al., *Application of Data Mining Techniques in the Analysis of Indoor Hygrothermal Conditions*, SpringerBriefs in Applied Sciences and Technology, DOI 10.1007/978-3-319-22294-3_4

Fig. 4.1 Buildings layout

corresponding to typical use conditions. In all units, pressurization and depressurization tests were performed. One year of continuous monitoring of temperature and relative humidity was planned. To measure the two parameters, data loggers Model HOBO U12-011 and HOBO UX100-011 were used. Temperature measurements in these devices ensures a values range between −20 and 70 °C, with ±0.35 °C of precision and 0.03 °C resolution. The recording relative humidity ensures a spectrum measurement values between 5 and 95 % a precision ±2.50 % and a resolution of 0.03 %, according to the manufacturer.

The installation of the temperature and relative humidity loggers in the different flats was established according to ISO 7726 (ISO 1998). The sensors were placed near an interior wall to prevent the combined effect of solar radiation and the influence of outdoor air temperature, which most likely would cause a distortion of the final results. They were installed at a height of about 1.10 m, ensuring that the sensors are protected from: possible abnormal cooling due to possible high external obstacles, air infiltration due to open doors or windows, and convective processes that can occur in the bathrooms.

4.3 Measurement Results

The field campaign produced extensive results that were not meant to be explored in detail in this text. Rather, as an example, 1 week in the heating season and 1 week on the cooling season, were selected to present the variability found in the measurements of each apartment. Although, in general, two rooms were monitored in each apartment, only the result in each bedroom was focused in this analysis. The dwellings' airtightness was also measured, leading to an average Rph50 of 8.9 h^{-1} with a coefficient of variation of 35 %, minimum value of 3.8 h^{-1} and maximum value of 15.1 h^{-1}. This means that the dwellings, although geometrically similar, presented relevant differences regarding airtightness due to window maintenance and user actions. The overall values allow to classify the dwellings as leaky.

4.3.1 Outdoor Conditions

The outdoor temperature daily patterns over 8 days in January and August are displayed in Fig. 4.2. Graphically it is possible to notice the different temperature variation patterns in 8 consecutive days. To make a more complete characterization of the outdoor conditions the same analysis is made for the exterior vapour pressure. Figure 4.3 presents the exterior pressure values of 8 consecutive days in typical weeks of January and August. The results are in line with the Koppen-Geiger classification of the Portuguese climatic conditions, which in this case correspond to Csb, dry-summer, presenting moderate temperatures and rainy weather due to the Atlantic Ocean influence. This type of climate brings about a specific user behaviour towards the building. Since the exterior conditions aren't extremely cold during winter, the users will often reduce heating to a minimum and try to balance economic savings with minimum comfort conditions. During summer, on the other hand, cooling is not often needed and hence the dwellings do not usually have cooling equipments installed.

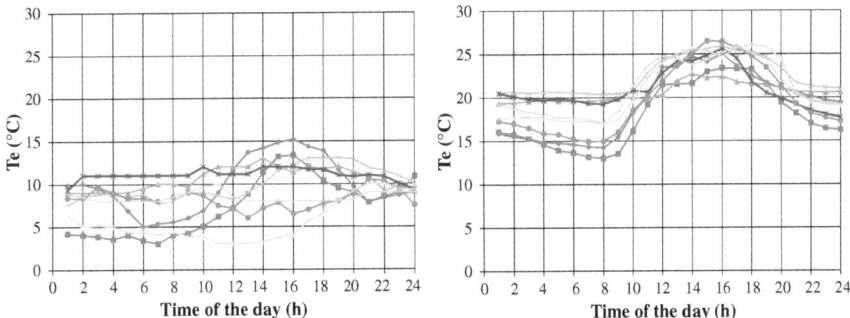

Fig. 4.2 Outdoor temperature through 8 consecutive days in January and August

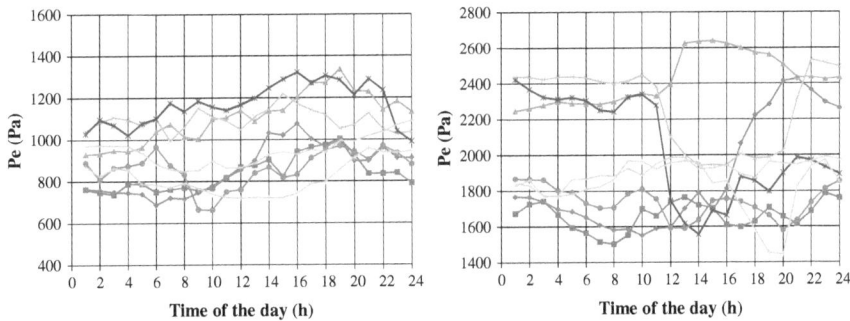

Fig. 4.3 Outdoor vapour pressure through 8 consecutive days in January and August

4.3.2   Indoor Temperature

In Figs. 4.4 and 4.5 it is possible to observe the variation of the hourly average indoor temperature for the different flats main bedrooms. It is found that average temperatures present higher temperature range in the winter than in the summer season. Also, although continuous heating is not common due to economic reasons, it's still possible to find dwellings where that will occur. When dealing with different flats, due to specific conditions (heated/unheated, number of inhabitants, window frame, cooking activity…) it may be difficult to find a pattern that allows a global description of the sample. Nevertheless, summer conditions were actually quite similar for all the flats, proving that the pattern observed in winter will not necessarily be observed in summer.

   In the case of winter conditions, the variation of both mean values between dwellings and peak to mean difference in each dwelling presented strong variations. Not only was the magnitude of the daily amplitude highly variable but also the period for peak occurrence would vary as well. These initial results, associated to the homogeneity of the dwelling sample highlighted the importance of user behaviour especially in winter conditions, where the loads and ventilation habits become the main cause for the observed variability.

   The comfort conditions for the week presented as example are also interesting to notice. The majority of the dwellings presented temperatures consistently below 15 °C, which imply low thermal comfort for the users. On the other hand, overheating during summer was not observed, putting the analysis focus on winter conditions.

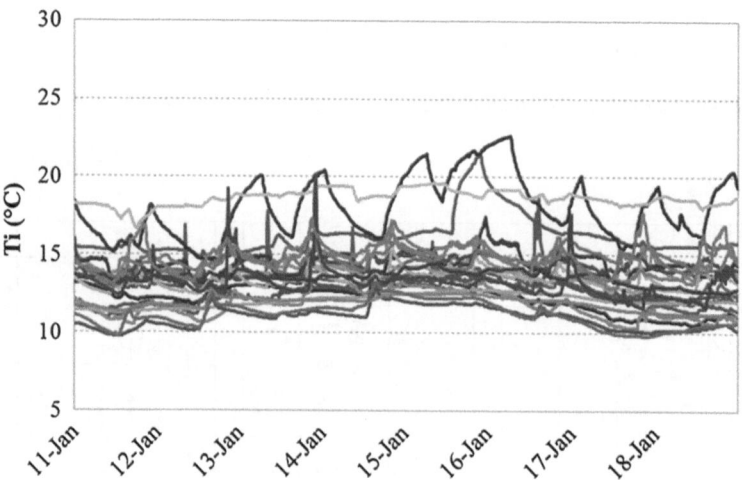

Fig. 4.4 Indoor temperature through 8 consecutive days in January in each flat of the sample

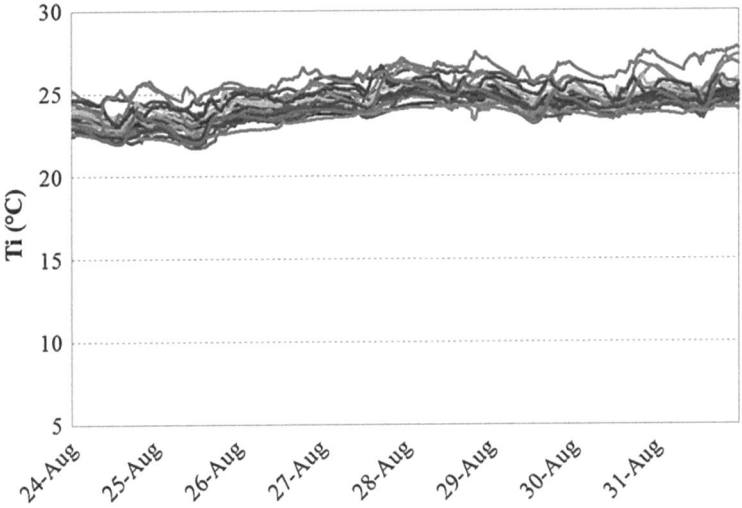

Fig. 4.5 Indoor temperature through 8 consecutive days in August in each flat of the sample

4.3.3   Indoor Relative Humidity

In Figs. 4.6 and 4.7 the hourly average indoor relative humidity values in the different flats is presented. As the data from the temperature, the average relative humidity presented a higher variation range in the winter than in the summer

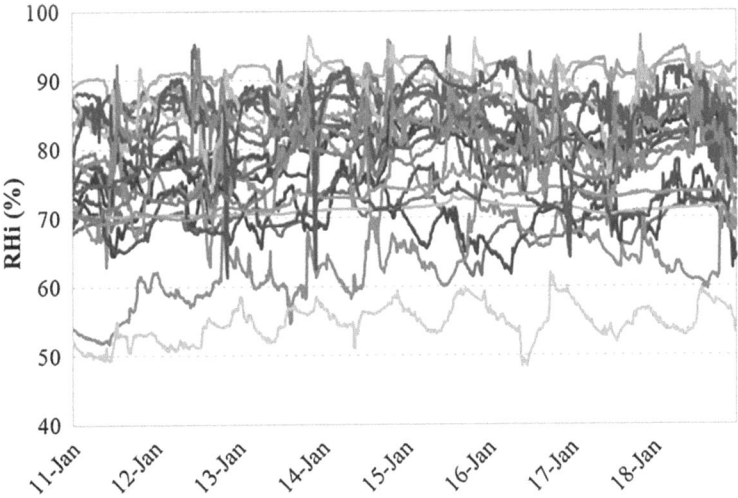

Fig. 4.6 Indoor relative humidity through 8 consecutive days in January in each flat of the sample

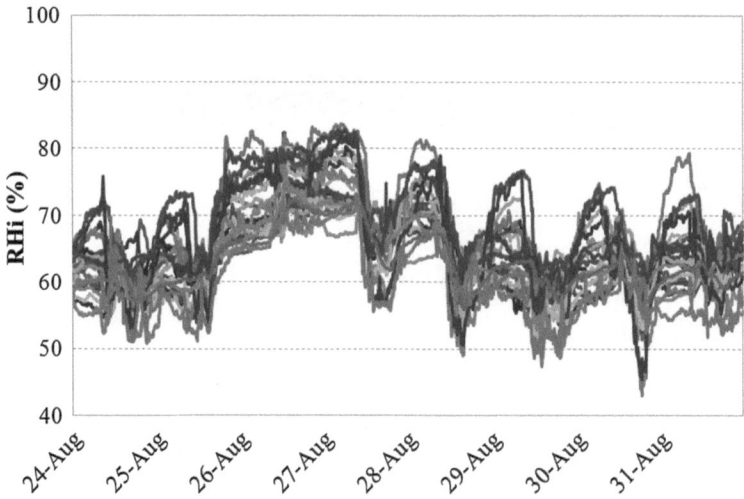

Fig. 4.7 Indoor relative humidity through 8 consecutive days in August in each flat of the sample

although the difference is rather less significant. What is also interesting is the greater difficult for defining patterns of relative humidity variation.

The analysed dwellings had small areas, implying small volumes and hence a fast relative humidity variation due to user action. This shows that temperature and relative humidity may have independent patterns, along with the differences between seasons. The values found also show that the risk of mould growth will be relevant, given the fact that relative humidity values were above 80 % in a substantial part of the dwellings. Mould growth associated to low thermal comfort may implies additional health hazards that should be investigated for this type of specific context.

Even a small analysis period as this one proved that analysing indoor hygrothermal conditions cannot be done easily. Although this sample was quite homogeneous, with similar buildings, occupied by users with the same economic context, still very different conditions were observed.

References

CEN EN 13829 (2000). *Thermal performance of buildings—determination of air permeability of buildings—fan pressurization method*. Brussels, Belgium: European Committee for Standardization.

ISO ISO 7726 (1998). *Ergonomics of the thermal environment—instruments for measuring physical quantities*. Geneva, Switzerland: International Organization for Standardization.

Chapter 5
Application of Data Mining Techniques

Abstract In this chapter, the data mining techniques described in chapter 3 are applied to the case study described in chapter 4. The overcome of the difficulties associated with the analysis process of the data using appropriate techniques is demonstrated.

5.1 Statistical Analysis of Data

The hygrothermal characteristics of the 24 apartments, indoor temperature and relative humidity, as well as the airtightness, net floor area and number of people, were the variables used to perform the statistical study. Regarding the hygrothermal parameters, the daily values of January in the different flats are analysed.

Regarding the characteristics of houses' occupancy, the variables "Area" and "Number of people", the frequency table is described in Table 5.1. It turns out that the apartments have areas ranging from 49 to 75 m^2 and the number of inhabitants per apartment varies between 1 and 5. However, the sample distribution reveals that the apartments are mostly occupied by only two persons.

Concerning the scale of the daily average indoor temperature (Ti), it is possible to observe a temperature range from 10.87 to 18.52 °C, with a relatively homogeneous distribution of the sample, a mean of 13.64 °C and a standard deviation of 1.89 °C. Concerning the daily indoor relative humidity (Hri), it is possible to observe a minimum value of 53.33 % and a maximum value of 94.55 %, with a sample mean equals to 78.64 % and a standard deviation of 8.76 %. The airtightness (Rph50) of the flats varies between 3.5 h^{-1} and 15.4 h^{-1}, with a mean of 8.86 h^{-1}. These and other descriptive measures of the variables under study are presented in Table 5.2.

The analysis of histograms and boxplots related to the variables daily average indoor temperature, daily average indoor relative humidity and airtightness, are presented in Figs. 5.1, 5.2 and 5.3, respectively. These Figures show that the distributions of these variables are symmetrical. Additionally, it is possible to observe that there are no outliers other than a home that has an average indoor

© The Author(s) 2016

N.M.M. Ramos et al., *Application of Data Mining Techniques in the Analysis of Indoor Hygrothermal Conditions*, SpringerBriefs in Applied Sciences and Technology, DOI 10.1007/978-3-319-22294-3_5

Table 5.1 Two-way frequency table of area and number of people per apartment

		Area (m²)					Total
		49	52	60	72	75	
No people	1	0	0	0	0	1	1
	2	2	5	4	1	1	13
	3	1	0	0	2	0	3
	4	0	1	1	0	2	4
	5	0	0	0	2	1	3
Total		3	6	5	5	5	24

Table 5.2 Descriptive statistics

Descriptive statistics		Ti (°C)	Hr (%)	Rph50 (h⁻¹)
Mean		13.64	78.64	8.86
95 % confidence interval for mean	Lower bound	12.84	74.94	7.50
	Upper bound	14.44	82.33	10.22
5 % trimmed mean		13.53	79.07	8.80
Median		13.56	78.81	8.70
Variance		3.57	76.68	10.34
Std. deviation		1.89	8.76	3.22
Minimum		10.87	53.33	3.50
Maximum		18.52	94.55	15.40
Range		7.65	41.22	11.90
Interquartile range		2.34	14.06	3.95
Skewness		0.84	−0.87	0.37
Kurtosis		0.79	1.68	−0.46

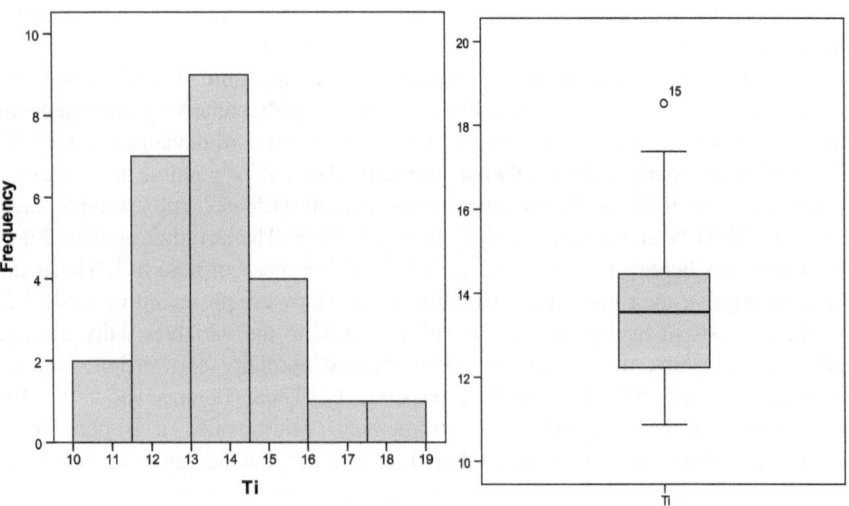

Fig. 5.1 Histogram and boxplot of indoor temperature

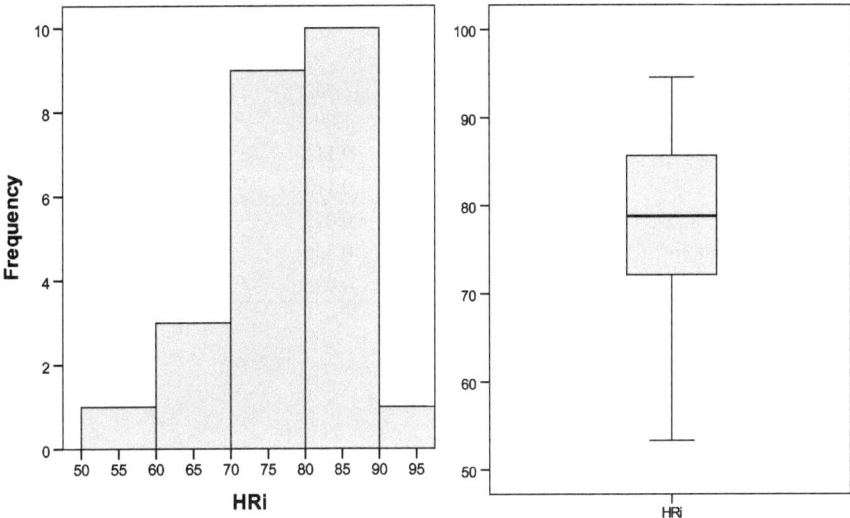

Fig. 5.2 Histogram and boxplot of indoor relative humidity

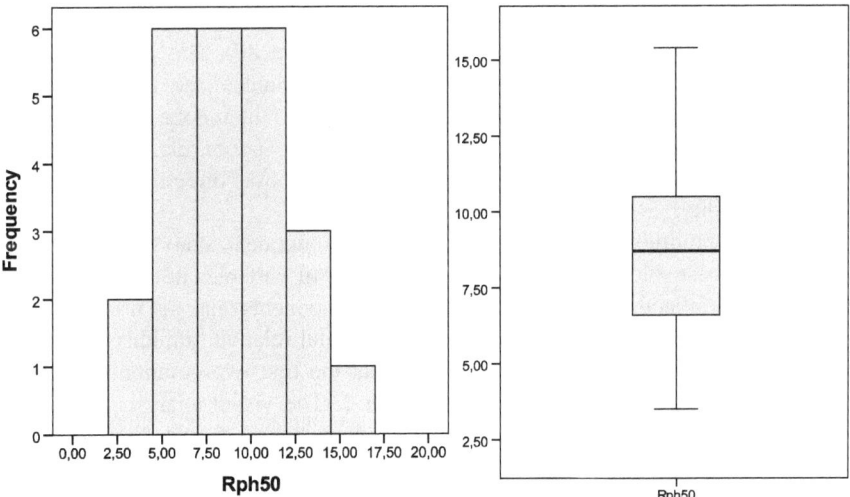

Fig. 5.3 Histogram and boxplot of Rph50

temperature above the others, which indicates that it is a house with continuous heating.

The symmetry and shape of these distributions point to a normal distribution of these variables, so the statistical Kolmogorov-Smirnov test was applied to verify it. The test results are presented in Table 5.3, showing that these variables are normally distributed to a significance level of 5 %, since p-value >5 %.

Table 5.3 Tests of normality

		Ti	HRi	Rph50
Normal parameters	Mean	13.64	78.64	8.86
	Std. deviation	1.89	8.76	3.22
Most extreme differences	Absolute	0.112	0.132	0.103
	Positive	0.112	0.092	0.103
	Negative	−0.072	−0.132	−0.089
Kolmogorov-Smirnov Z		0.548	0.647	0.505
p_value		0.925	0.797	0.960

5.2   Multivariate Data Analysis

For the multivariate analysis of the data 11 variables were used: 4 related to indoor temperature, 4 related to the indoor relative humidity, the area (Area), the number of people (N_people) and the airtightness (Rph50). To characterize the internal temperature were used: the daily average (Ti), the 10 % percentile (Ti_10), the 90 % percentile (Ti_90) and the period in which the temperature was below 18 °C (T < 18). To characterize the indoor relative humidity were used: the daily average (Rhi), the 10 % percentile (Rhi_10), the 90 % percentile (Rhi_90) and the period in which the relive humidity was higher than 80 % (Rhi > 80). The study started by analysing the possible correlations between the 11 variables (see Table 5.4). This preliminary analysis showed that the variables related to the indoor temperature are strongly correlated, as well as the variables related to the indoor relative humidity. It was also noted that the number of people inside the houses directly influences the relative humidity.

After this preliminary analysis, the principal component analysis was used to establish the best solution for reducing the number of variables in order to include only the uncorrelated ones. Thus, the principal components analysis was applied to the 8 variables related to the indoor temperature and relative humidity. From this analysis (see Table 5.5) should be retained that the first two components are the most relevant with an eigenvalue greater than 1. The visual analysis of the plot screen (see Fig. 5.4) withdraws the same conclusion as should be selected all components until the line that unites them start getting horizontal, that is, to submit a reduced slope.

Applying again the principal component analysis selecting now two components (see Table 5.6), it appears that the two components together explain 95.239 % of the total data variability. The first component explains 64.519 % and the second 30.720 %. The positive, and close to 1, Cronbach's alpha value for both components suggests that these components are reliable.

Table 5.7 describes the variance of each of the original variables explained in the new major components. The representation allows realizing which one, or ones, are the most important variables for determining each of the major components. Thus, all variables are associated with their 1st component, but are also associated with

Table 5.4 Correlation matrix

	N_people	Area	Rph50	Ti	Ti_10	Ti_90	Ti < 18	HRi	Hri_10	Hri_90	Hri > 80
N_people	1.000	0.334	−0.018	0.194	0.159	0.179	−0.161	**0.459**	0.363	**0.561**	**0.479**
Area		1.000	0.154	0.246	0.232	0.194	−0.190	0.020	0.049	0.051	0.049
Rph50			1.000	0.171	0.236	0.148	−0.177	−0.148	−0.092	−0.209	−0.224
Ti				1.000	**0.966**	**0.986**	**−0.976**	−0.199	−0.239	−0.206	−0.298
Ti_10					1.000	**0.933**	**−0.941**	−0.223	−0.236	−0.257	−0.337
Ti_90						1.000	**−0.956**	−0.165	−0.200	−0.171	−0.271
Ti < 18							1.000	0.117	0.177	0.132	0
HRi								1.000	**0.961**	**0.969**	**0.917**
Hri_10									1.000	**0.897**	**0.844**
Hri_90										1.000	**0.944**
Hri > 80											1.000

Table 5.5 Model summary

Component	Cronbach's alpha	Variance accounted for	
		Total (Eigenvalue)	% of variance
1	0.922	5.181	64.766
2	0.669	2.412	30.147
3	−7.125	0.138	1.728
4	−7.721	0.129	1.612
5	−15.939	0.067	0.836
6	−31.462	0.035	0.438
7	−49.438	0.023	0.282
8	−73.852	0.015	0.190
Total	1.000	8.000	100.000

Fig. 5.4 Principal components

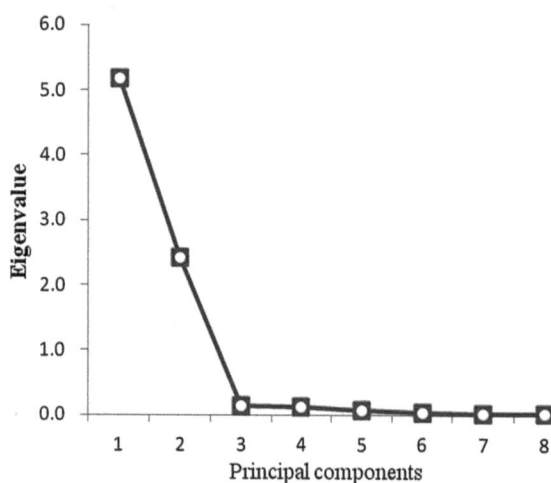

Table 5.6 Model summary

Component	Cronbach's alpha	Variance accounted for	
		Total (Eigenvalue)	% of variance
1	0.921	5.162	64.519
2	0.678	2.458	30.720
Total	0.993	7.619	95.239

the 2nd component. This same conclusion is also demonstrated by the component loadings presented in Table 5.8. From the analysis of component loadings one can state that the variables related to the relative humidity are opposed to the variables related to temperature.

Figure 5.5 shows the component loadings and indicates the realization of each dwelling on the main components.

Table 5.7 Variance accounted for

	Total (vector coordinates)		Total
	Component		
	1	2	
Ti	0.675	0.306	0.981
Ti_10	0.710	0.240	0.951
Ti_90	0.703	0.242	0.944
HRi	0.677	0.298	0.976
Hri_10	0.620	0.309	0.929
Hri_90	0.640	0.318	0.958
Ti < 18	0.620	0.291	0.911
Hri > 80	0.517	0.452	0.969
Active Total	5.162	2.458	7.619
% of Variance	64.519	30.720	95.239

Table 5.8 Component loadings

	Component loadings	
	1	2
Ti	0.822	−0.553
Ti_10	0.843	−0.490
Ti_90	0.838	−0.492
HRi	−0.823	−0.546
Hri_10	−0.787	−0.556
Hri_90	−0.800	−0.564
Ti < 18	−0.787	0.540
Hri > 80	−0.719	−0.673

Fig. 5.5 Component loadings

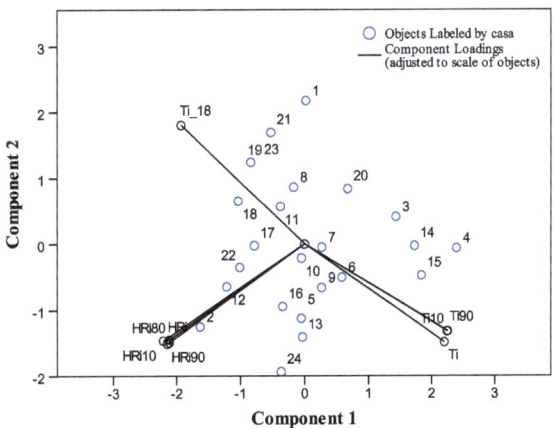

After this first analysis, the reduced and better organized data can now be used as input for applying the cluster technique for a criteria-based grouping of the houses. The goal of applying cluster analysis in this research is to identify the homogenous groups of houses, in other words those having a similar performance, and afterwards using this information to identify possible improvement strategies. The appropriate variables that were selected for clustering are: components 1 and 2, area, number of people and airtightness. Since these variables are measured on different scales, those with larger values would contribute more to the absolute value of the distance than the ones with smaller values. Therefore, in this study, as in several other examples, prior to cluster analysis, all variables were first normalized to the interval 0–1.

In this example, the agglomerative procedure has been used. The dissimilarity in the consumption between each pair of dwellings is measured by the normalized Euclidean distance (or straight-line distance). The distance between clusters, or groups of houses in this example case, can be quantified by different methods. These methods differ in the way the difference (distance) between the clusters is accounted for. An interesting fact is that when applying different methods, different clustering results can arise. The average linkage and the Ward's linkage clustering methods were applied to calculate the distance between the different clusters of houses.

A common way to visualize the cluster analysis progress is by drawing a dendrogram that displays the distance level at which there was a combination of objects and clusters. This kind of graphs are a very simple tool that can be used to extract useful information about the sample properties and concerning the most important variables for the clustering final result. The dendograms of the clustering results applying both the Ward's method and the average linkage method to the houses characteristics are presented in Fig. 5.6.

An important problem in the application of cluster analysis is the decision regarding how many clusters should be derived from the data. As presented in Fig. 5.6 both clustering methods allow us to identify 2 clusters among the 24 houses. Cluster 1 includes houses 1, 2, 3, 4, 5, 14, 15, 16, 17 and 18 and cluster 2 contains the remaining houses. From the results some immediate findings can be pointed as the houses of the cluster 1 are bigger (areas between 72 and 75 m^2) than the houses of cluster 2 (areas between 49 and 60 m^2). This means that the houses would first be distinguished due to their size even after the parameter normalization.

If another option is taken, considering the vertical cut sooner, different clusters can be identified would arise from the two methods. With this approach, the new clusters would reflect the influence of hygrothermal conditions as well as the dwellings' size, confirming that the difference of the calculation methods implies

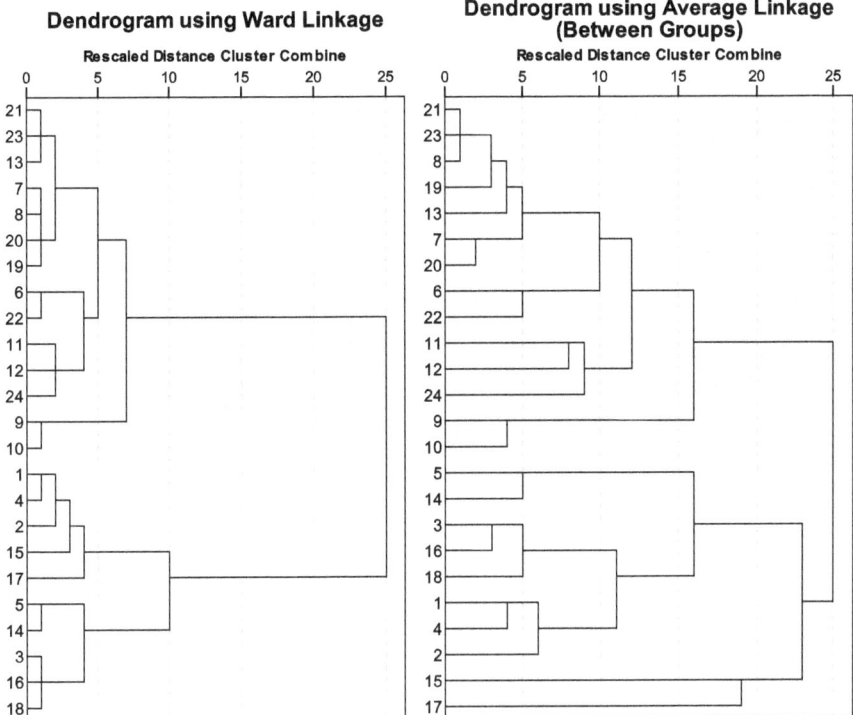

Fig. 5.6 Dendograms of clustering houses by using Ward's method (*left*) and average linkage (*right*)

differences in the cluster composition. The sensitivity of Ward's method to outliers' identification is clear as in this method dwellings 15 and 17 would be separated from the rest of the houses. These dwellings had prior revealed a clearly different behavior.

Chapter 6
Conclusions

Abstract This chapter presents the main conclusions, stressing how descriptive statistics combined with multivariate analysis can be applied to a specific case study.

The possibilities created by modern monitoring techniques of gathering extensive amounts of hygrothermal indoor measurements created the need for advanced analysis techniques application to extract useful information. In this work, Data Mining was the pursued strategy, especially for finding hidden patterns in the extensive available data.

The analysis of indoor hygrothermal conditions has been targeted by a wide number of researchers. The concept of indoor environmental quality is very broad and depends on many variables. Yet, thermal comfort is unanimously recognised as crucial for an adequate indoor environmental quality and, in particular, air temperature and relative humidity are the more common parameters selected for long time monitoring and several simplified models only use these two parameters. Along with the time series that result from in situ measurements, key performance indicators are also frequently calculated to better characterize the building's consumption and performance patterns.

Data mining includes many different techniques. In this work, a clear division is made between basic statistical tools and multivariate data techniques. The basic statistical tools application starts by descriptive statistics to provide basic information about variables and highlight potential relationships between variables. After that first step, where key characteristics are defined through summary statistics, probability distributions should be derived. Correlation matrices are an important tool for searching relationships in the data.

The multivariate data techniques can start by performing a principal components and factor analysis. This may be decisive to reduce the number of variables considered in the analysis. This technique allows to summarize the information in a smaller number of variables called factors/main components. Cluster analysis can be applied to define the groups (clusters) within a data set, following a two step approach. Different methods can be used to perform cluster analysis but, in this work, special attention was given to hierarchical agglomerative methods, in which

N.M.M. Ramos et al., *Application of Data Mining Techniques in the Analysis of Indoor Hygrothermal Conditions*, SpringerBriefs in Applied Sciences and Technology, DOI 10.1007/978-3-319-22294-3_6

subjects start in separate clusters that are successively combined until all subjects are included in the same cluster.

A case study was introduced to allow for an example application of the data mining techniques. A set of 24 flats belonging to the same neighbourhood were monitored for a long period to assess their indoor hygrothermal behaviour, including time series of Temperature and Relative humidity, along with specific parameters such as airtightness. The winter conditions displayed a higher variability and hence the month of January results were selected for the application of data mining techniques. Also winter season proved to be frequently uncomfortable and associated to the risk of mould growth. The very leaky dwellings with low income occupants that try to use the minimum as possible of heating energy explain those conditions.

The statistical study of the 24 apartments was performed for the month of January, using indoor temperature and relative humidity time series along with airtightness, net floor area and number of occupants. Net floor area ranged from 49 to 75 m^2 and the number of inhabitants from 1 to 5, with most of the dwellings having usually two occupants. The analysis of histograms and box-plots showed that the daily average indoor temperature and relative humidity, as well as air-tightness, presented a symmetrical distribution. The Kolmogorov-Smirnov test proved that the variables were normally distributed. Eleven variables were used in the multivariate analysis. To characterize the internal temperature were used: the daily average (Ti), the 10 % percentile (Ti_10), the 90 % percentile (Ti_90) and the period in which the temperature was below 18 °C (T < 18). To characterize the indoor relative humidity were used: the daily average (Rhi), the 10 % percentile (Rhi_10), the 90 % percentile (Rhi_90) and the period in which the relive humidity was higher than 80 % (Rhi > 80). This analysis showed that the variables related to the indoor temperature are strongly correlated, as well as the variables related to the indoor relative humidity. It was also noted that the number of people inside the houses directly influences the relative humidity. The principal components analysis was applied to the 8 variables related to the indoor temperature and relative humidity. The first two components were the most relevant with an eigenvalue greater than 1. From the analysis of component loadings one can state that the variables related to the relative humidity are opposed to the variables related to temperature.

The cluster analysis was applied considering as variables components 1 and 2, area, number of occupants and airtightness. Ward's method and average linkage were applied and the resulting dendograms were analysed. One important conclusion is that the interpretation of the cluster analysis results should be cautious since, depending on the vertical cut option, different clusters may be found.

Initially, the statistical measures and graph illustrations allow us to analyze and interpret the data, as well as explain their variations. The data mining techniques applied have led to the identification of two groups of houses, in which, after a cursory inspection indicates that its area is a major factor.